The Landscapes of Craters of the Moon National Monument

The Landscapes of Craters of the Moon National Monument

An Evaluation of Environmental Changes

R. Gerald Wright

and

Stephen C. Bunting

Northwest Naturalist Books   University of Idaho Press   Moscow, Idaho   1994

Copyright © 1994 R. Gerald Wright and Stephen C. Bunting

University of Idaho Press, Moscow, Idaho 83844-1107

Printed in the United States of America.

All rights reserved.

No part of this publication may be reproduced, stored in a retrieval system, or transmitted in any form or by any means, electronic, mechanical, photocopying, recording, or otherwise, except for purposes of scholarly review, without the prior permission of the copyright owner.

Design by Karla Roberts

98 97 96 95 94   5 4 3 2 1

Library of Congress Cataloging-in-Publication Data

Wright, R. Gerald.
 The landscapes of Craters of the Moon National Monument : an evaluation of environmental changes / R. Gerald Wright and Stephen C. Bunting.
   p. cm. — (Northwest naturalist books)
 Includes bibliographical references and index.
 ISBN 0-89301-168-1 : $24.95
 1. Vegetation dynamics—Idaho—Craters of the Moon National Monument. 2. Botany—Idaho—Craters of the Moon National Monument—Ecology. 3. Ecology—Idaho—Craters of the Moon National Monument. 5. Craters of the Moon National Monument (Idaho) 6. Vegetation dynamics—Idaho—Craters of the Moon National Monument—Pictorial works. 7. Botany—Idaho—Craters of the Moon National Monument—Ecology—Pictorial works. 8. Ecology—Idaho—Craters of the Moon National Monument—Pictorial works. 9. Natural history—Idaho—Craters of the Moon National Monument—Pictorial works. 10. Craters of the Moon National Monument (Idaho)—Pictorial works.
 I. Bunting, Stephen C. II. Title. III. Series.
 QK156.W75 1994
 581.5'265'0979659—dc20   93-34720
                                                    CIP

Contents

Foreword
vii

Introduction
1

The Environment of
Craters of the Moon National Monument
History and Exploration — Geology — Soils — Climate — Vegetation — Visitor Use
3

The Concept of Environmental Change
11

Long-term Evaluation of Environmental Change
Using Matched Photos to Evaluate Environmental Change — Photographs Used in This Study
13

Vegetation Change on the Snake River Plain
19

The Photographs
21

What the Photo Comparisons Reveal
91

Appendix
Common and Scientific Names of Plants
97

Literature Cited
99

Index
101

Figures

Figure 1
Craters of the Moon National Monument
4

Figure 2
Craters of the Moon National Monument
Photo Locations
9

Figure 3
Photo Locations
Northern Part of Craters of the Moon National Monument
13

Figure 4
Photo Locations
Middle Part of Craters of the Moon National Monument
14

Figure 5
Photo Locations
Southern Part of Craters of the Moon National Monument
16

Foreword

Over the years we have spent many delightful hours traveling throughout the Snake River Plain. Generally these trips have been in connection with different research studies or as part of long-term efforts to examine plots or representative plant communities selected to monitor ecological change. A common ingredient of our travels has been a penchant, fortunately equally shared, to travel routes theretofore unvisited, and along these routes to explore the unknowns and attempt to answer ecological questions these new landscapes revealed. Our wanderlust has resulted in the fact that, between the two of us, there now are few roads in the region that at least one of us has not driven. Our desires for a better ecological understanding of passing phenomena often resulted in the abrupt stopping of the vehicle to try and identify a new plant or to discern the reasons for the existence of a particular geologic formation or plant community. Fortunately again, our mutual curiosities were self-reinforcing—and many trips, even between two relatively close locations, could turn into all-day excursions.

We are convinced our travels have made us better ecologists. However, we have also come to realize that our curiosities and quests for knowledge were not simply fortuitous but were in fact a learned behavior. They were derived in part from experiences gained through our long relationship, both as a friend and mentor, with Dr. Edwin Tisdale, now Professor Emeritus in the Range Resources Department at the University of Idaho. Ed was a member of the faculty of the Range Resources Department for most of his career, retiring in 1987. Even after retirement he remained active in the pursuit of his research projects, most of which focused on central and southern Idaho. Ed is a true fountain of knowledge with respect to the grasslands and shrublands of the West. He taught many specific facts; but more important, he taught us to ask questions and to seek to understand the causality underlying natural effects. He also reinforced our own ideas about the value of natural, undisturbed areas. It is for these reasons we dedicate this book to Ed Tisdale.

This book is an extension of our curiosities about those factors which have and continue to shape environment of the Snake River Plain. We have probably interpreted only a few of them. Others may see things we have missed or doubt some of our interpretations. The beauty of the medium of photography is that it easily permits such insights, and we welcome such comments. The sites we have selected are protected and can be easily visited, either now or in the future, thus permitting someone to do their own interpretation of what the landscape reveals. In addition to furthering individual knowledge of the ecology of Craters of the Moon and of the concept of environmental change, we hope this work inspires others to look at their own photo collections in a different light and to use them to evaluate the changes which might have taken place since their photos were originally taken.

Peggy Pace, Ed Tisdale, and Harley Wright reviewed the manuscript; their suggestions have been invaluable—correcting areas of fact and interpretation. The different research projects at Craters of the Moon which provided us the opportunity to undertake this project were funded by the National Park Service through the Pacific Northwest Regional Office. The staff of Craters of the Moon National Monument has always, over the years, been a joy to work with. They facilitated the logistics of many of our projects, provided us with access to all photo files and, most important, provided friendship, support, and encouragement. In this regard, special credit goes to Neil King and Dave Clark.

Introduction

Although often unrecognized, changes in the natural environment are a normal part of our world. Some changes are a consequence of natural factors over which humans have little control; others may be the direct or indirect consequence of human intervention. Some components of the environment change more rapidly than others. The effects of some changes, such as the erosion of river canyons, are so subtle they can only be detected over long periods of time. Others, such as revegetation following fire, are detectable in decades or less. Any determination of the magnitude of environmental change requires a standard or baseline against which to measure the resulting alterations.

Records providing baseline or average conditions are relatively common for environmental changes that fluctuate daily or seasonally, such as climate changes and the volume of stream flow. Baseline measures for environmental conditions which do not change as frequently are less common. One reason is that most environmental research studies are designed to answer specific questions and they are not usually designed to be repeated in later years, a situation which would provide information on changes over time in the resources studied. Attempts to replicate older studies are usually complicated by the fact that the methods are often poorly explained or statistically unsound, and original plots and original data may be lost.

There is, however, one form of data which, where available, are reliable and relatively easy to use and which provide a basis to help explain and examine changes in resources and landscapes over time. These are identical pairs of historic and contemporary photographs of a given area which span a time interval long enough to record the environmental changes of interest.

In this book, we have sought to provide, through the use of comparative photographs, illustrations of how a natural landscape can change over time. To permit a focus on natural ecological changes, we have chosen a region which has been protected from most forms of human development and resource manipulation. As the reader will see, some components of this landscape have remained remarkably stable while others show considerable alteration even over relatively short time periods. Most of the changes we highlight are not dramatic; rather they are often subtle alterations in vegetation and landform which occur continuously in nature as a result of a variety of environmental factors.

We conducted this study at Craters of the Moon National Monument, which is located on the northern edge of the Snake River Plain in southern Idaho. The monument is administered by the U.S. National Park Service (NPS). As with other national parks and monuments, most forms of consumptive resource uses, such as mining, the harvest of timber, grazing by domestic livestock, and hunting are prohibited at Craters of the Moon. In addition, motorized access is greatly restricted and facility development is limited to specific locations. These conditions allow the changes caused by natural environmental factors to be separated from those caused by human actions.

There is a tendency to view all environmental change as being detrimental. It is true that many of the environmental changes that are observed, even in isolated areas, are caused by increasing intrusions of pollutants, the introduction of exotic species, and changing land uses in adjacent areas. However, other changes are quite natural and may be a consequence of changes in climate or in the dynamics of animal populations. Still other changes induced by humans may be viewed as beneficial, as when formerly disturbed sites are rehabilitated to their natural condition.

Our purpose in producing this book is in part to raise the public's awareness of the dynamics of the natural environment and the changes which can and do occur in nature, and in part to inform also those individuals who manage the natural environment. Management decisions for some natural areas are often made in the absence of sufficient ecological data and in a milieu of intense public scrutiny. On a growing number of environmental issues there is a plethora of individual opinions about what the best management strategies should be. We feel that an understanding of environmental change can provide invaluable insight and benefit the process of making sound and justifiable judgments about managing natural resources.

The photographs compiled in this book are the result of 12 years of joint study on a variety of resource concerns at Craters of the Moon National Monument and throughout the Snake River Plain of southern Idaho. During this period we located and took over 120 comparative photographs. Those included here are most representative of the spectrum of ecological conditions at Craters of the Moon National Monument.

Typically viewed by passing visitors as a harsh, sterile, and unforgiving place, many people would probably concur with one of the first views of the area published 156 years ago by Washington Irving who wrote:

> The volcanic plain . . . forms an area of about sixty miles in diameter, where nothing meets the eye but a desolate and awful waste; where no grass grows nor water runs, and where nothing is to be seen but lava. Ranges of mountains skirt this plain, and, in Captain Bonneville's opinion, were formerly connected, until rent asunder by some convulsion of nature. Far to the east, the Three Tetons lift their heads sublimely, and dominate this wide sea of lava;—one of the most striking features of a wilderness where every thing seems on a scale of stern and simple grandeur.

Almost prophetically he noted: "We look forward with impatience for some able geologist to explore this sublime, but almost unknown region" (Rees and Sandy 1977, 98).

The real Craters of the Moon is, however, very much the opposite of this, offering an incredibly diverse array of biotic resources and a variety of fascinating landforms. This diversity is also presented on a scale small enough to allow an observer to gain an intimate understanding of the place. During the course of our studies we have developed a special attachment for Craters. We hope that through the photos and text of this book others can gain this same appreciation for the landscape. One facet this book cannot convey, however, is the sense of solitude and peace of mind one attains at Craters of the Moon; to gain this one has to visit the monument.

The Environment of Craters of the Moon National Monument

History and Exploration

The first documented exploration of the periphery of the Craters of the Moon area was led by Benjamin L. E. Bonneville, an army officer and fur trader who extended the search for fur beyond the Snake River in 1833-34 (Ostrogorsky 1983). During the 1840s and 1850s, thousands of emigrants crossed the Snake River Plain along the Oregon Trail. In 1862 in search of new routes, some emigrants left the Snake River at Fort Hall and took Goodale's Cutoff, which ran along the northern edge of the Snake River Plain and skirted the Craters of the Moon lava field. A portion of Goodale's Cutoff passes through the north end of the monument. Diaries of the emigrants vividly describe what by 1904 pioneers referred to as Craters of the Moon. The unpleasant experiences of the first pioneers with the harsh environment of the lava fields caused later settlers to avoid the area. As a result, the monument remained largely unexplored until the early 1900s.

Stearns (1928) reported that Arco resident J. W. Powell searched for water supplies for livestock in the Craters of the Moon lava field in 1879 and again, with Walter Ferris, in the 1880s. I. C. Russell of the U.S. Geological Survey (USGS) led the first scientific exploration of the northern Craters of the Moon area as part of his broader studies of the Snake River Plain in 1902, but he avoided the central portion of the monument. Harold T. Stearns, also of the USGS, undertook the first of his many explorations of the area in 1921 (Stearns 1924). This initial exploration was followed by the expeditions led by Robert Limbert, a Boise adventurer, photographer, and writer. Limbert's account of the expedition and his dramatic photographs were published in National Geographic Magazine in 1924 (Limbert 1924). Limbert's article and the studies conducted by Stearns were the first to recognize the unique geological value of the area and were, along with the personal testimony by these explorers, largely responsible for the proclamation by President Calvin Coolidge establishing the monument in 1924. Coolidge's proclamation was made to protect the unique series of volcanic cones, craters, lava flows, and caves located along the Great Rift. Additional lands were added in 1928 to protect the watershed north of the park in the foothills of the Pioneer Mountains.

A second addition was made in 1962 to add Carey Kipuka located southwest of the boundary of the monument. A kipuka is an area of older landscape surrounded by recent volcanic flows. The 180-acre Carey Kipuka was considered to be particularly unique and valuable as an ecological baseline because there was no evidence that it had ever been grazed by domestic livestock and it was covered by the mature sagebrush-grass vegetation that was once common in the region but which has since been significantly reduced by fire, grazing, agriculture, and competition from exotic species (Tisdale et al. 1965).

Craters of the Moon National Monument now encompasses 54,000 acres with elevations ranging from 4,940 to 7,160 feet. A map of the monument with major facilities and features identified is shown in Figure 1. In 1970, approximately 80 percent of the monument, the area south of the road

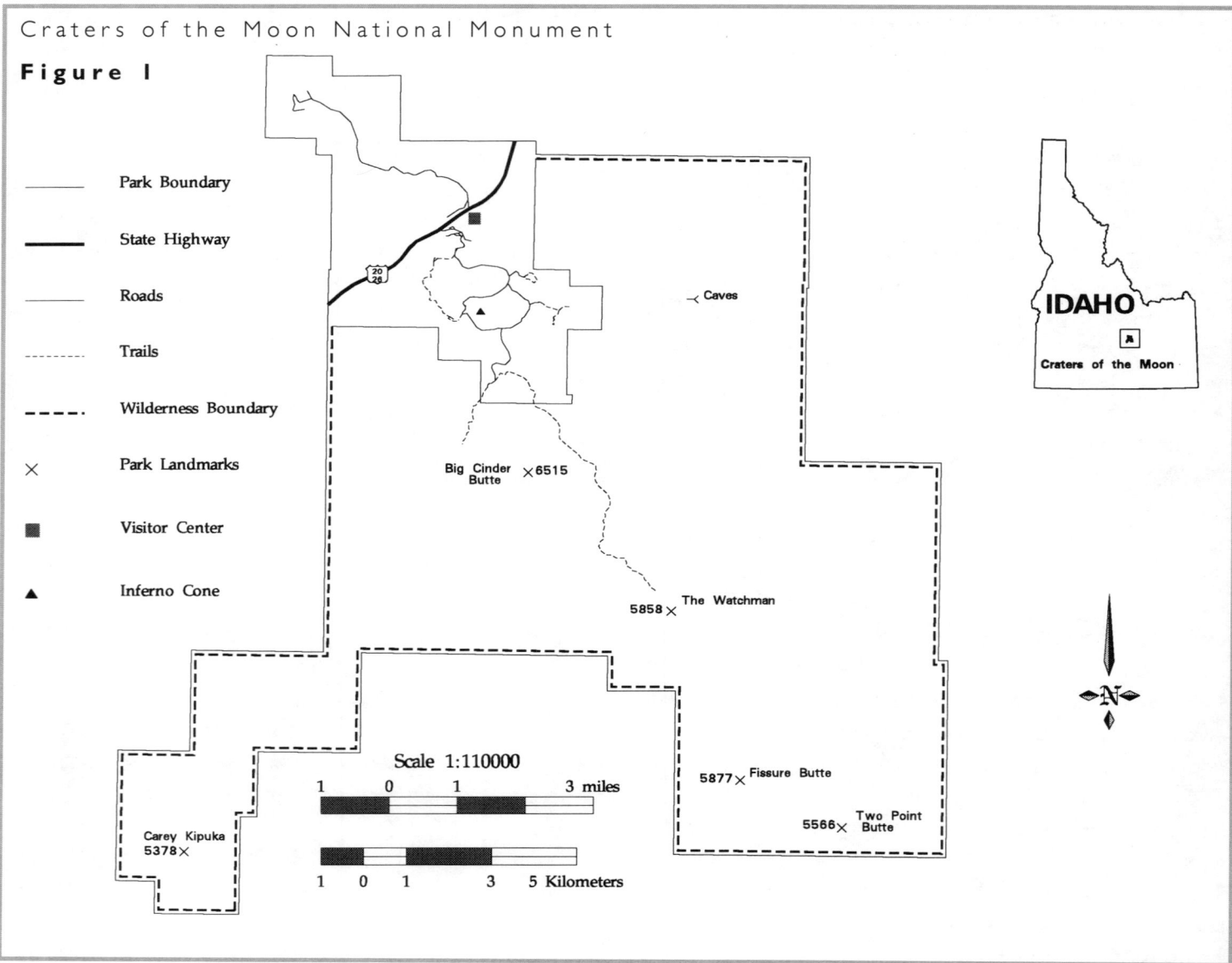

system, was legislatively classified as wilderness. This classification precludes all motorized access and assures that human disturbance will be minimal.

Geology

The Great Rift is a 55-mile long north to northwest trending fissure that may be likened to a great elongated volcano without a central orifice (Stearns 1963). Craters of the Moon lies approximately 15 miles west of what was once the most active portion of the Great Rift, and it is characterized by a series of northwest aligned cinder cones, craters, and spatter ramparts (Prinz 1970).

The earliest published estimate of the age of the Craters lava flows is Russell's (1902). Because of the lava's fresh appearance and the size of the limber pines, Russell judged the most recent flows to be no more than 100 to 150 years old. Stearns (1924) also was impressed by the apparent youth of the lava flows, but noted that one would expect the amount of vegetation on flows in the arid climate of the Snake River Plain to be less than in volcanic areas of similar age in the humid climate of Hawaii. Stearns counted 461 rings in a core from a tree growing on a pahoehoe flow near Big Craters and estimated that the most recent eruptions had occurred 250 to 1,000 years previously. After a core taken in 1954 from the Triple Twist Tree showed that it was at least 1,350 years old, Stearns (1963) revised his estimate of the age of the most recent lava to at least 1,650 years.

There have been extensive studies over the past 20 years by scientists from the USGS to better understand the evolution of the Craters of the Moon lava field (see Kuntz et al. 1986a). These studies have concluded that lava flows of the Craters lava field were set down during at least eight eruptive periods that began about 15,000 years ago and ended about 2,000 years ago. Each eruptive period lasted no more than several hundred years. The intervals between eruptive periods on the lava field ranged from several hundred to 3,000 years and averaged about 2,000 years. This sequence has led scientists to predict that "because the present interval has lasted 2,200 years, another eruptive period seems likely to occur within the next 1,000 years" (Kuntz et al. 1986b, 174). In addition, because past eruptions at the Craters of the Moon lava field have generally occurred in areas of the Great Rift that have been quiescent the longest, these scientists predicted the next eruption will begin in the central part of the monument near Big Cinder Butte and, possibly, will move to the northern part of the Great Rift.

Only one type of lava is known to occur in the Craters lava field. This is a dark colored rock rich in iron and poor in silica called basalt. It covers the Snake River Plain and forms the rimrock along the Snake River. There are two general types of lava flows on the monument. The Hawaiian words aa and pahoehoe have been adopted as scientific terms to describe these flows.

Pahoehoe, which covers about half the area of the monument, is a billowy, ropy type of lava that has numerous caverns. Its shiny blue glassy crusts make some of the flows beautiful in bright sunlight. The ropy and wrinkled surfaces of the pahoehoe are due to the hardening of a thin crust or scum on the lava flow while the crust is being pushed forward by the flowing lava below. The caves that are found at Craters of the Moon all occur in the pahoehoe lava. They were formed within the flow itself by the hardening of the surface of the lava stream. As the flow continued the side walls also stiffened and a tube was formed. The tube conducted lava without much loss of heat to the advancing margin of the flow. When the flow ceased, lava continued to drain out of the upper portion of the tube eventually leaving an empty cavern (Stearns 1963).

Aa lava has jagged surfaces with sharp points, which makes it extremely difficult and somewhat dangerous to walk across. The heat and gas content of aa lava is different from pahoehoe. During an eruption, gas escaping from the doughy lava mass pulled out stringers of lava which subsequently cooled causing the spines and jagged surface of the flow. There are aa lava flows 25 to 100 feet thick on the monument, some of which have moved several miles out on the plain.

Three different types of volcanic structures occur on the monument—cinder cones, spatter cones, and lava domes. The cinder cones have black, loose, cindery surfaces and smooth conical profiles. They represent the heaps of lava froth or spray formed from the fire-fountains of the eruptions. Big Cinder Butte is considered to be one of the largest purely basaltic cinder cones in the world (Stearns 1928). Many of the cinder cones are elongated to the northeast, presumably the result of prevailing southwest winds at the time of their formation. The spatter cones were formed by smaller fire-fountains. The clots of lava hurled out by these fountains were not sufficiently inflated with gas to form cinders and they traveled such a short distance that they fell as clots in a viscous state and adhered to one another.

Lava domes consist of compact lava and have a broad, flat dome shape, many rising only 30 to 50 feet above the surrounding country. These inconspicuous domes are formed by the continuous eruption of pahoehoe lava on the surface from the same point. Because most of the lava escapes through tubes the cone is not built up very high. Examples of lava domes include Indian Tunnel and Great Owl Cavern.

Soils

The soils of Craters of the Moon are derived primarily from basaltic magma or cinder material resulting from the different eruptive periods (Blakesley and Wright 1988). On the older flows, the soils have developed from loess or residuum—common soil parent materials on the Snake River Plains and adjacent foothills. The soil structure of the cinder cones and adjoining areas is very difficult to characterize because of the presence of buried soil profiles. This phenomena occurred when the fallout from a newer eruption covered older, more mature surface soils that were developing an organic layer and may have been covered with vegetation. In some areas of the cinder cones, several buried surface horizons can be encountered which in turn influence nutrient and moisture availability and thus vegetation patterns (Day and Wright 1989).

Climate

The climate of the monument is typical of the upper elevations of the Intermountain Sagebrush/Steppe region. Average annual precipitation at monument headquarters is about 18 inches. The annual precipitation has a bimodal pattern with a relatively high peak in December in the form of snow and a smaller peak in May in the form of rain. A period of relative drought occurs from mid-June through September. Temperatures below freezing have been recorded in all months at the monument at least once over the past 25 years.

Vegetation

Over half the monument consists of relatively barren lava flows. The vegetated areas are dominated by sagebrush/steppe communities intermixed in many areas with limber pine. Sagebrush/steppe communities are a mixture of different species of sagebrush, antelope bitterbrush, rubber rabbitbrush, wax current, and perennial grasses such as bluebunch wheatgrass and Idaho fescue. Douglas-fir occurs on the north-facing slopes of older cinder cones and in the foothills of the Pioneer Mountains to the north. Many of the vegetated areas are separated from surrounding areas on the Snake River Plain by barren lava flows up to three miles wide. These flows make travel difficult and have contributed to the region's isolation. They have also restricted use by domestic livestock prior to the monument's establishment.

The scarcity of free water over most of the monument also has prevented significant use by domestic livestock and still limits mule deer use of the southern two-thirds of the monument during late summer (Griffith 1983). Mule deer generally use this portion of the monument only up through the time when their requirements for water can be met by the moisture in the vegetation that is eaten. After July, the drying of the vegetation limits the available moisture and necessitates that the deer move to the northern portion of the monument where perennial streams provide needed water. All of the above factors, combined with the management protection afforded by the National Park Service and the area's legislated wilderness status, have resulted in little non-natural disturbance to the plant communities, and many areas appear to be in pristine condition.

The vegetation of Craters of the Moon is highly diverse and ranges from simple communities with a small number of total species on recent lava and cinder material to well-developed, complex Douglas-fir and sagebrush communities. A diverse riparian community that includes dense aspen stands is found in the northern section of the monument. Day and Wright (1985) described 26 distinct vegetation community types within the monument. The majority of the vegetation of the monument is in the early stages of primary succession, e.g., it is on recent volcanic material or in the transitional shrub/limber pine types and hence is not well developed. Monument plant communities have thus been excluded from published descriptions of common habitat types for southern Idaho (Steele et al. 1981, Hironaka et al. 1983). The vegetation on the well-developed soils of the Pioneer Mountain foothills is included in one of these habitat type classifications.

Visitor Use

Over the past decade, an average of 215,000 people per year have visited the monument. Most visitors stay only a few hours and tour the 5.6-mile loop road stopping at various interpretive sites, although during the summer months the 52-unit campground is full on most nights. The most popular attractions are the spatter cones and the several lava caves that are open to the public on the trail system. The spring wild flower display on the cinder cones is an event of special importance drawing many visitors. In recent years winter visitation has increased dramatically coincident with efforts by the monument to make the roads available for cross-country skiing.

Craters of the Moon National Monument—Photo Point Locations

Figure 2

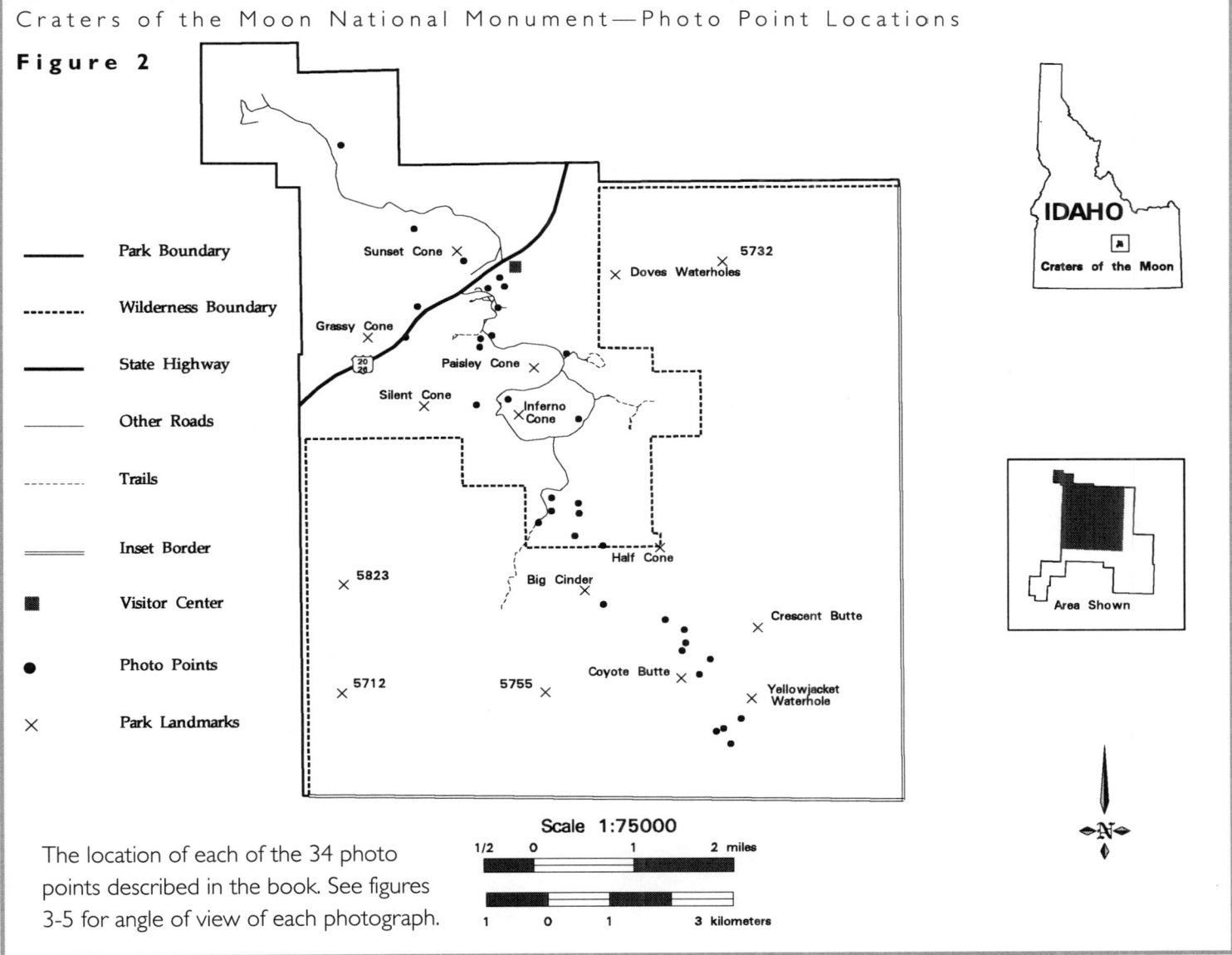

The location of each of the 34 photo points described in the book. See figures 3-5 for angle of view of each photograph.

The Concept of Environmental Change

Changes in the environment can occur over widely differing time frames and alterations may vary substantially in the effects they have on the environment. Some changes occur at geologic time scales measured in tens of thousands of years (10^4 years). They include phenomena like continental drift, mountain uplift, and continental glaciation. Because these events encompass such large spatial areas, e.g., often whole continents, and persist so long, often thousands of years, they are difficult to study and comprehend. A different magnitude of environmental change is caused by phenomena like volcanic eruptions, earthquakes, hurricanes, and floods, which typically occur at time scales measured in centuries (10^3-10^2 years). These events involve much smaller geographic areas and are of relatively short duration, although their impacts may persist for decades. These events are also difficult to study because their frequency of occurrence and location are unpredictable. At the other end of the time spectrum are changes in the environment which occur relatively frequently, e.g., on a daily, monthly, or yearly basis (10^{-2}-10^0 years). These include changes in herbaceous plant biomass, changes in the densities of annual plants, and the rhythms of daily physiological cycles. The timing and predictability of these events make them easier to study and to understand. For this reason, events which occur in this time scale have been the focus of most traditional ecological research.

The time frame we focus on in this book lies inbetween the above extremes. This is the time scale of decades (10^1 years). We believe that many of the environmental changes which occur within this time scale often go unnoticed because of their subtle nature and because the long-term programs needed to study them have not been conducted. This time frame has been referred to as "the invisible present" because the kinds of environmental changes which occur within it represent one of the most critical gaps in ecological research and, therefore, in our ability to understand the environment (Magnuson 1990). The kinds of alterations which can occur within this time frame and which are illustrated in the photographs include changes in plant community structure and composition caused by variations in climate, fires, or fire suppression, and the invasion of non-native plants.

Unlike changes which occur at longer time intervals such as those associated with earthquakes and hurricanes, the events which occur within the invisible present are far easier to predict and manage. It is therefore ironic they are not better understood. The reason most applied ecological research seems to encompass a relatively short time period is because most studies originate in response to management problems that require a quick solution. Conversely, long-term research studies are relatively uncommon in the natural sciences. Only the collection of climate and stream-flow data have received much long-term emphasis (Mack et al. 1983).

One of the reasons why long-term studies to monitor environmental change are not common is that there is little glamour in collecting most data: it is repetitive and systematic work. As a result, this task is often relegated to temporary employees who by the nature of their positions lack the perspective to see its value. Administrators see little payoff from these programs since they do not solve immediate "problems." Finally, it is often difficult if not impossible to obtain funds to collect routine data as opposed to conducting basic research involving new enviromental information (Wright 1992). Administrators of most land management agencies have often failed to understand the fact that environments change, and therefore have been opposed to funding routine monitoring studies. In addition, even where they have been established, routine monitoring programs are often the first to face budget cuts. Land management agencies like the National Park Service are only now beginning to recognize that long-term research and monitoring of environmental change are important in developing an understanding of park ecosystems (Swanson and Sparks 1990).

Long-term Evaluation of Environmental Change

There are essentially two approaches to gaining an understanding of long-term environmental change. The first is to establish an ecologically sound and statistically valid resource monitoring program in a given location and to systematically continue this program for many years. There is growing interest in the use of parks as "benchmarks" or examples of undisturbed areas against which to compare the impacts of environmental change in surrounding areas. With the increasing development of North America, this concept is gaining prominence and funding. Unfortunately, since most of the studies established under this type of program are of recent vintage, it may be many years before interpretable results are produced (Silsbee and Peterson 1991).

A second approach is to incorporate previously collected information into an on-going program in order to give it a long-term time reference. Unfortunately, as alluded to previously, such studies are difficult to replicate because original plot locations were not identified or are now lost, and others are difficult to interpret because data were not systematically collected. The use of comparative historic and contemporary photographs avoid or mitigate many of these pitfalls, and comparative photographs can be very useful in examining and interpreting environmental changes that have occurred at a particular site.

Using Matched Photos to Evaluate Environmental Change

When properly used, matched pairs of historic and contemporary photographs can provide a clear record of the changes that have occurred at a particular site over the intervening time period. They have proven particularly useful for evaluating changes in vegetation structure and species composition which have occurred on a given site (e.g., Christensen 1957, Hastings and Turner 1965, Gruell 1980, 1983, Houston 1982, and Rogers 1982).

To the authors' knowledge, none of the studies cited above or others in the literature have used photographs that were originally taken with the express purpose of providing a future baseline of environmental change. Most studies, including this one, have relied almost exclusively on photographs taken by explorers who sought only to document unique aspects of the scenery or natural resources or social conditions along the route of travel.

Previous investigators using comparative photographs have concentrated on a specific location with a history of old photographs like Yellowstone National Park (Houston 1982) or used photographs from organized explorations (e.g., the W. H. Jackson photos from the Hayden survey [Gruell 1980]). Craters of the Moon does not have a history of photography and was not on the route of the famous western explorers. For this book we were therefore limited to photographs taken by early explorers and by park personnel over the years. Most of the latter were taken primarily to document changes or planned changes in park development. During the course of our studies we perused several hundred photographs. As might be expected, most were not useful in interpreting environmental change, and only that subset which met our objectives (which was usually incidental to the objective of the original photographer!) were incorporated in our study.

The methods for matching old photographs have been described by Harrison (1974) and Rogers (1982). The usual technique is to have the old photograph in hand, and to initiate a search for the exact location of the picture through a trial and error process until the scene best matches the original. The search process can be very time consuming when, as was the case with many old photographs, only a general location, if any, is provided with the original. It can also be very satisfying, particularly when an original scene emerges and becomes "painfully obvious" after hours of searching. The search process is usually complicated when there are no identifiable foreground objects in the old picture. Framing a duplicate is made more difficult by the fact that in most instances the camera type and the lens focal length used in the original are not known. The time of year and time of day can also affect the ability to accurately match the original photo. Complications in interpretation can arise from different patterns of shadows when the time of day that the original photograph was taken is unknown, a factor which is normally the case (Rogers 1982). In our study, we had no knowledge of the time of day of any of the original photographs, and even lacked specific knowledge of the day, month, and year on some of the original photos. Where known, we did match the month of the original whenever possible.

Photographs Used in This Study

All of the 35 original photographs included in this study are the property of the NPS. Nine of the original photos were taken by Robert Limbert during his 1923 exploration of Craters of the Moon. Most of the photographs and original negatives from Limbert's explorations were never published and were lost for many years. About 75 nitrate-based photographic negatives taken by Limbert during the exploration were discovered in 1982 by historian Michael Ostrogorsky, who was working with author Gerald Wright to write a biography of Limbert for Craters of the Moon National Monument. He found the negatives stored in the Boise, Idaho, attic of an elderly relative of Limbert. The negatives were eventually donated to the monument, and they are now cataloged and preserved by the Idaho State Historical Society. The majority of Limbert's photographs were of geological features and of the activities of the expedition. We selected only a small subset from the originals that adequately depicted scenes of the plant communities of Craters of the Moon.

Nine of the original photos were taken by D. S. Scofield, an NPS engineer who was doing a study in 1935 to design a road system for the monument. Most of the proposed roads extended south of the existing network into the present Wilderness Area and were never constructed. Seven originals were taken by NPS biologist Coleman Newman during a

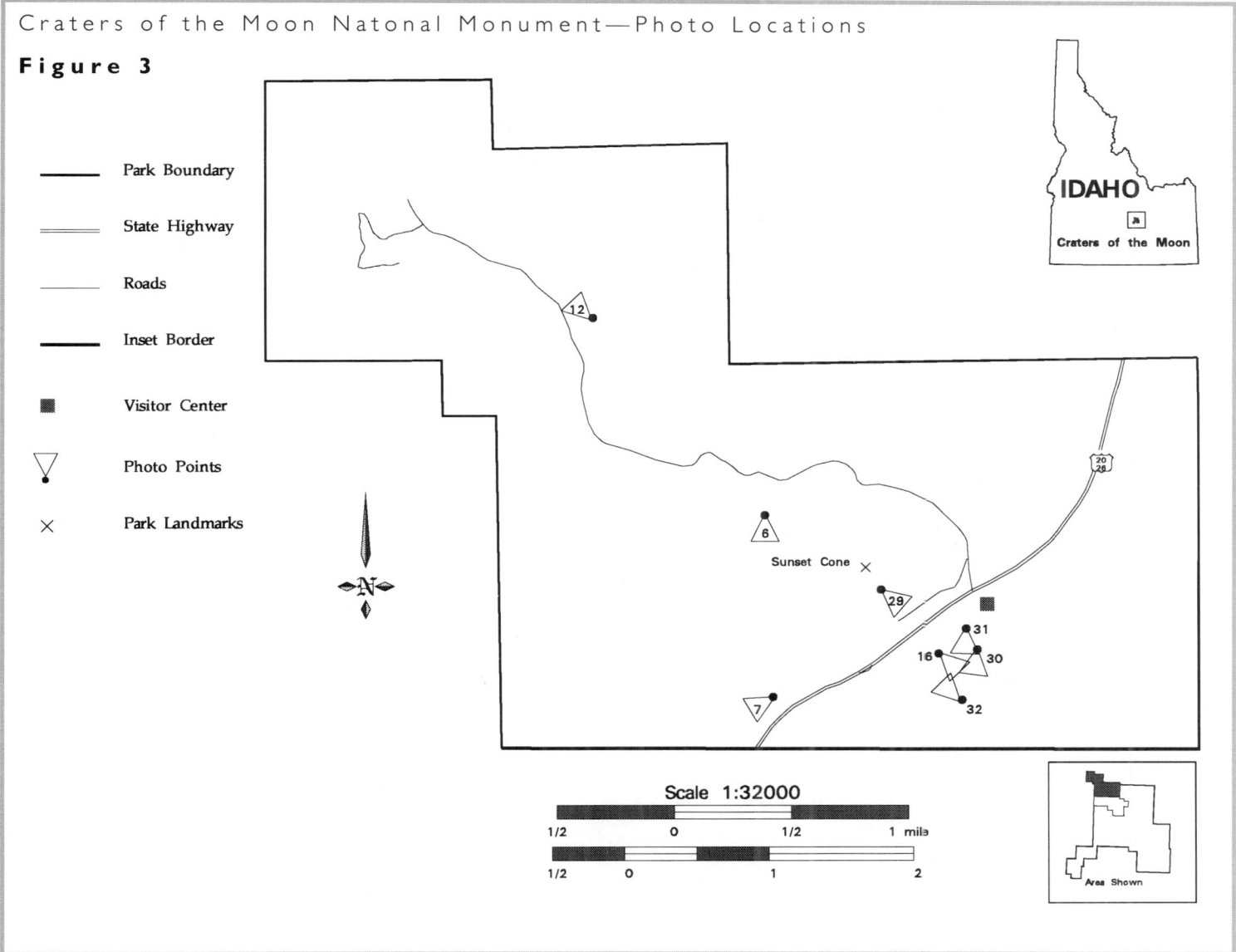

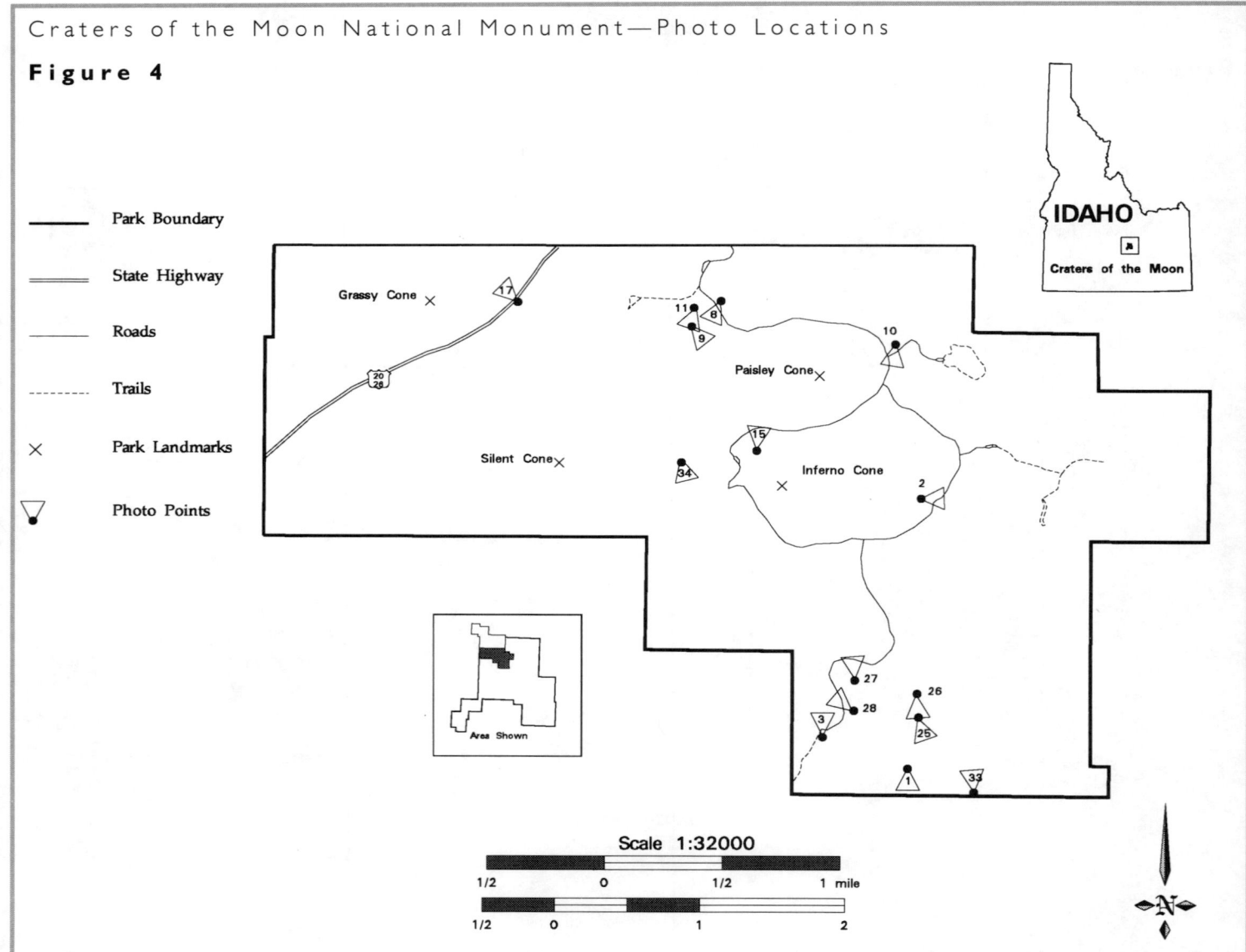

Craters of the Moon National Monument—Photo Locations

Figure 5

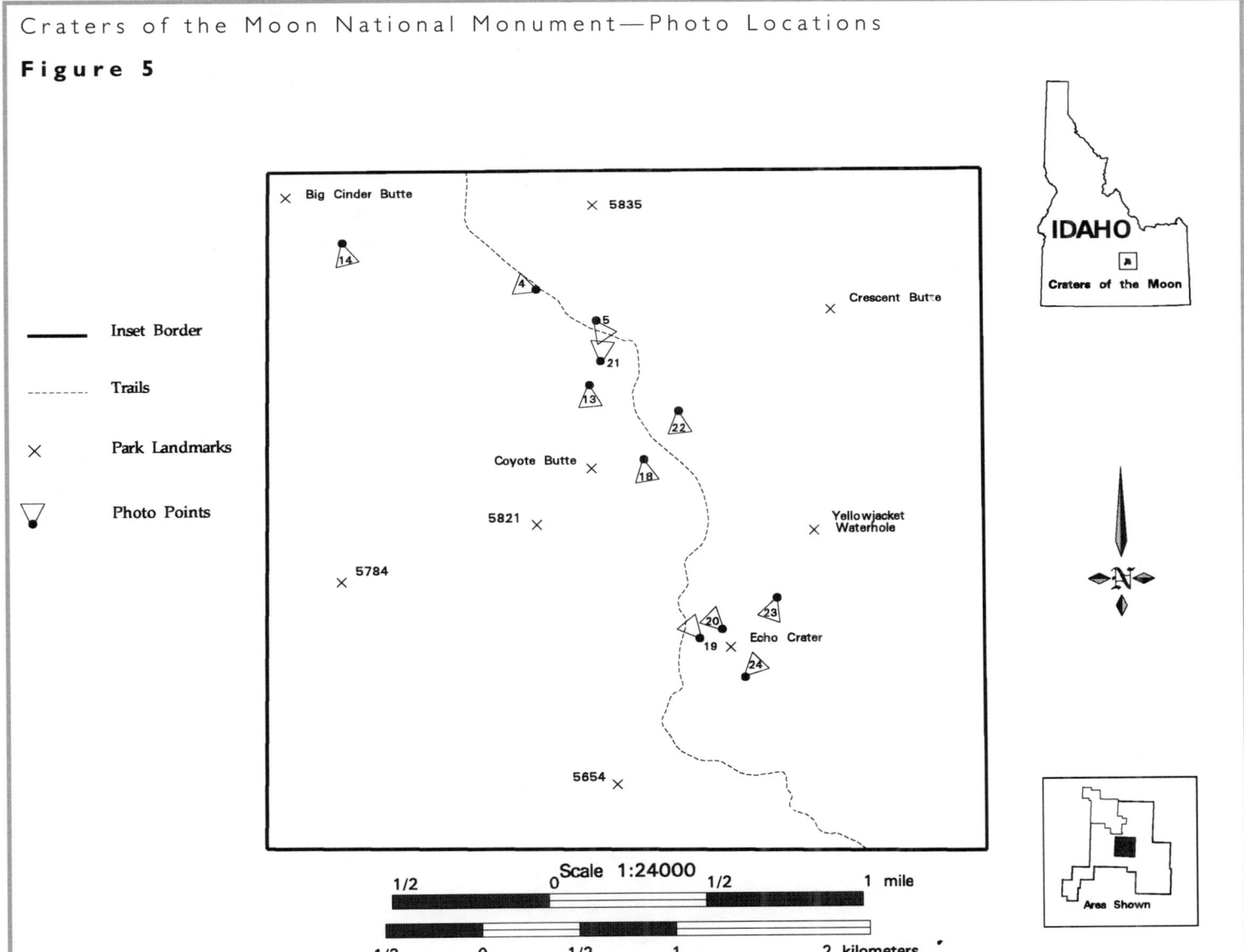

brief survey of the monument's biological resources in 1957. The majority of the remaining originals were taken by NPS personnel stationed at the monument and were obtained from the files and archives of the monument. Approximately 25 percent of the photos in this book were taken in the 1920s, 35 percent in the 1930s, 27 percent in the 1950s, and 13 percent in the 1960s. The most recent comparative photographs were taken in 1965. There is a lack of original photos from the 1940s. This probably reflects the scarcity of film during the war years and the diminished staffing due to severe cuts in personnel and funds.

All of the contemporary photos were taken by the authors during the spring and summer months between 1986 and 1990. Detailed documentation of the camera station location, camera type, film, and exposure used was made for each new photo. An overview of the location of each of the photo points in the monument is shown in Figure 2. Figures 3 through 5 show the specific locations of the photo points with their respective fields of view.

Vegetation Change on the Snake River Plain

Several previous studies have discussed historic changes in the vegetation of the Snake River Plain of southern Idaho and the Bonneville Basin of north-central Utah. All reveal a similar sequence of events. They have assumed that prior to settlement by Euro-Americans, the vegetation in the region consisted largely of mixed-aged stands of sagebrush/steppe. Fire was an occasional occurrence but did not greatly change the plant communities. Sagebrush might be diminished on a given site by fire, but over time it was gradually re-established from seed. Grasses and other herbaceous plants temporarily increased until dominance of sagebrush returned. Thus, although different sites had a different mosaic of vegetation as a result of short-term environmental changes, over the entire Snake River Plain the system was in a broad equilibrium.

Euro-American settlement in the mid 1800s altered this equilibrium. Settlement brought about increased grazing pressure by domestic animals and a reduction of the juniper woodlands because the trees were used for fuel and timber. By the late 1800s the result was a reduction in native perennial grasses due to overgrazing and possible declines in available water. The latter was caused by compacted soils resulting from the trampling of grazing animals, which produced greater run-off and less moisture infiltration into the soils. This in turn caused a further reduction in grasses and herbaceous plants and therefore reduced the occurrence and spread of fire. Consequently there was a rather steady increase in the densities of trees and shrubs, which has continued on some sites to the present time.

Over the long-term there has been a trend on drier sites, where competition from native grasses has been reduced, for a dramatic long-term increase in the densities of annual grasses. Many of these annual grasses are non-native and have, over time, been accidentally introduced onto the Snake River Plain. Cheatgrass is the most prevalent. It became common on the Snake River Plain between 1915 and 1920 (Mack 1981), and its presence has modified both fire occurrence and plant competition relationships in sagebrush/steppe vegetation. Cheatgrass sprouts and grows early in the spring, thereby taking advantage of early spring moisture before the later sprouting perennial grasses. By the time the native grasses are starting to put on seed, cheatgrass has already completed its life cycle and it has dropped its seed and the plant is cured out. Thus in the hot summer season it burns easily, allowing fires to burn through stands of shrubs, trees, and perennial grasses. The fires eliminate competition allowing even more cheatgrass to grow the following year. What is created is a positive feed-back cycle of increased densities of cheatgrass, more frequent fires, greater declines in native plants, and more opportunities for invasion by non-native species. As a result of this process, in many of the drier lowland areas of the Snake River Plain, the perennial grass and sagebrush species have been practically eliminated; and the sites are almost completely dominated by nonnative annual grasses which have little or no forage value.

Craters of the Moon is unique among most sites on the Snake River Plain in that it has been free of most of the above disturbances. The area within the monument was never settled and has, by all indications, received little historic livestock use. Since its establishment in 1924 it has not been grazed. Likewise, although the monument has had a fire suppression policy in effect over most of its history, major fires do not appear to have been common (Bunting and Wright 1991). Cheatgrass, while present throughout the monument, has not become dominant on most sites but rather has remained a minor component of the vegetation. Since the plant communities at Craters of the Moon have been free of the disturbances common elsewhere in the region, they offer a unique opportunity to examine biotic changes that are the result of natural ecological factors.

However, Craters of the Moon has not been totally immune from disturbance. Several of the photographs depict the changes brought about by the increase in roads and physical development necessary to accommodate growing numbers of visitors to the area. This has not, however, been a one-directional shift. Visitor use in terms of motor vehicle access is more controlled today than it was in the past, and some areas formerly intensively used have been restored to native conditions. In addition, all mechanized travel is prohibited in the 80 percent of the monument designated as wilderness in 1970.

Other changes have been brought about by NPS resource management activities. The most significant of these was the dwarf mistletoe control program of 1962 and 1963. At that time an outbreak of dwarf mistletoe in the limber pines of the monument was considered a threat not only to uninfected trees on the monument but also to tree stands in other areas. Even though dwarf mistletoe is native to the area, a decision was made to control the parasite by killing the infected trees either through poison or by cutting and burning them. Over the course of the program, park records indicate that about 6,000 trees were killed, most of them in the areas near the loop road in the northern end of the monument.

The Photographs

The 34 sets of photographs which follow illustrate a variety of vegetational changes which have occurred on the monument over the past 70 years. They indicate the dynamic nature of vegetation and the relative rapidity at which change may occur, even in severe environments.

The photographs have been arranged to illustrate the following themes: changes in limber pine density, changes in Douglas-fir density, changes in limber pine due to dwarf mistletoe control, changes in aspen abundance, the effects of fire, the constancy of plant distribution patterns in the shrub and cinder garden communities, the stability of the lava flows, and changes associated with park development. Following Gruell (1980), paired scenes have been numbered consecutively; the original, labeled "a," is on the left page and the retake, labeled "b," on the right. Captions under plate "a" describe the location of the photograph. Those under "b" describe the important relationships and changes that can be observed.

PLATE IA. **May 1923** The north slope of Big Cinder Butte, at 6,515 feet is the highest cinder cone in the monument and one of the largest purely balsaltic cinder cones in the world. It is uncertain exactly when in the sequence of eruptions at the monument this cone was formed. The group in the original picture are part of the Limbert exploration party. The photo point is on the lower southeastern slopes of the crater known as Broken Top. PHOTOGRAPH BY ROBERT LIMBERT

PLATE IB. **JUNE 1986 (63 years later)** There has been a substantial increase in the density of limber pine on the north slope of Big Cinder over the intervening time period. Most of the establishment occurs in the islands of shrubs which are primarily antelope bitterbrush, rubber rabbitbrush, and mountain big sagebrush, although some limber pine is being established on the bare cinder areas.

PLATE 2A. **Summer 1957** The camera faces northeast, positioned on the cinder flats southeast of Inferno Cone. The photo point is just off the loop road about 1/4 mile past the intersection of the road to the area occupied by lava structures known as tree molds. These are cone-like structures formed when hot lava surrounded the lower trunk of a tree burning the wood inside as the lava cooled around the trunk. NATIONAL PARK SERVICE PHOTOGRAPH, PHOTOGRAPHER UNKNOWN

PLATE 2B. **June 1987 (30 years later)** There has been a substantial (almost three-fold) increase in the number of limber pine present on these lower slopes. Shrub cover, consisting primarily of antelope bitterbrush, remains virtually unchanged. The level cinder-covered area in the foreground, referred to as cinder gardens by Day and Wright (1985), is covered primarily by dwarf buckwheat and silverleaf phacelia. It shows only minor evidence of change over the time period including the growth of a few shrubs in the foreground of the cinder gardens. Studies by Day and Wright (1985) have determined that the abrupt transitions in vegetation which occur between shrub zones and bare cinders are largely a consequence of soil moisture, which is influenced by buried soil horizons.

PLATE 3A. **Summer 1935** View looking due north across the rugged aa lava fields toward the Pioneer Mountains north of the monument boundary. The photo point is just south of the existing roadway at its entrance to the tree molds parking area. The camera position for the second photo is positioned somewhat higher because of grading done when the road was built. NATIONAL PARK SERVICE PHOTOGRAPH BY J. S. SCOFIELD

PLATE 3B. **June 1986 (51 years later)** Limber pine has increased in both size and density on the cinder-covered flats north of the extensive flow of pahoehoe lava in the foreground. Vegetation on this lava, which has virtually no soil cover, shows no apparent change.

PLATE 4A. **June 1957** View to the northwest of Big Cinder Butte. We could not precisely locate the original photo point or find the mounds of exposed lava rock. It was taken from the vicinity of the wilderness trail about two miles south of the tree molds parking area near a formation known as the lava trees. This area of the trail parallels the Great Rift and is known as Trench Mortar Flats. NATIONAL PARK SERVICE PHOTOGRAPH BY COLEMAN NEWMAN

PLATE 4B. **June 1986 (31 years later)** The numbers of limber pine trees have increased substantially over this time period, particularly in the draw between the south slope of Big Cinder and the adjoining hill. The broad barren area on the summit of Big Cinder remains almost entirely free of plant cover, probably because of soil factors and climate (wind and exposure).

PLATE 5A. **May 1923** This is a view of the southwest side of Crescent Butte, a large horseshoe-shaped cinder cone in the central portion of the monument. The photo point is just north of the wilderness trail in Trench Mortar Flats about 1,500 feet southeast of Plate 4. Although the wilderness trail did not exist at the time of Limbert's exploration, the topography and vegetation in this area probably dictated that he would have followed a path very similar to the present trail. PHOTOGRAPH BY ROBERT LIMBERT

PLATE 5B. **June 1988 (65 years later)** Limber pine numbers on the lower slopes of Crescent Butte have increased over the years. Increases in this area as opposed to the upper slopes may be attributable to greater down-slope soil moisture. The broad band of barren cinders transversing a portion of the butte remains free of vegetation, probably reflecting unfavorable subsoil characteristics. The lines of dwarf buckwheat on the cinder gardens in both photos are probably the result of depressions in the cinders which at certain times collect soil moisture and enhance plant establishment. These depressions were probably the result of mule deer tracks.

PLATE 6A. **Summer 1957** This view was taken from the second ridge northwest of the summit of Sunset Cone. Visible is the northwest lobe of the Highway Flow of aa lava and the north and northeast slopes of Grassy Cone and North Flow. NATIONAL PARK SERVICE PHOTOGRAPH BY COLEMAN NEWMAN

PLATE 6B. **June 1986 (29 years later)** The most obvious change is the marked increase in both numbers and canopy closure of Douglas-fir on the slopes of Grassy Cone. There is also a less marked but measurable increase in the number of Douglas-fir, which have spread out into the well-developed sagebrush/steepe surrounding the existing stand. The extensive sagebrush-dominated flats surrounding the lava of the Highway Flow have changed little if at all.

PLATE 7A. **July 1965** This view of the southeast slopes of Grassy Cone is from the photo point along the south edge of Highway 20 .7 miles west of the entrance road. This view shows the opposite side of Grassy Cone from that seen in Plate 6. NATIONAL PARK SERVICE PHOTOGRAPH BY ROGER CONTOR

PLATE 7B. **July 1986 (21 years later)** An increase in the canopy cover and density of Douglas-fir in the stand is evident as is the spread of a small number of new trees out into the sagebrush/steppe. Most Douglas-fir stands in the monument tend to be similar to this and occur in relatively dense stands with limited understory vegetation. In most cases, more than 50 percent of the soil surface is devoid of vegetation but is covered by a layer of litter. There is no evidence of change in the sparse vegetation of the cinder garden in the foreground.

PLATE 8A. **May 30, 1936** View of the east northeast slopes of North Crater with the loop road in the foreground. The photo point is about 50 yards east of the road and approximately .5 miles from the entrance station. National Park Service photograph by K. Dole.

PLATE 8B. **May 30, 1986 (50 years later)** It is estimated that approximately 90 percent of the mature limber pine trees on this section of North Crater were removed in 1962 and 1963 as part of the dwarf mistletoe control program. In this area, probably because of its proximity to the loop road and campground, the program involved cutting the trees and removing or burning the debris. The decrease in the density of limber pine is clear although the effect is mitigated by increases which have occurred in the 30 years since the removal program.

PLATE 9A. **Summer 1957** View of the northeast slopes of North Crater taken from the ridge above the North Crater parking area. The loop road is seen in the center of the picture. The photo point for Plate 9b is located in the lava field to the left of the truck in Plate 8b. NATIONAL PARK SERVICE PHOTOGRAPH, PHOTOGRAPHER UNKNOWN

PLATE 9B. **June 1987 (30 years later)**   This photo provides another glimpse of the large decrease in limber pine resulting from the mistletoe control program. Most of the trees now seen on the upper slopes have become established following the removal program. The distinct patterns of barren cinder areas remain unchanged despite the removal of the dominant overstory species which were not, at the time, growing on the barren cinder areas.

PLATE 10A. **Summer 1957** View of the southeast slopes of Paisley Cone. The intersection of the loop road with the turnoff to Devil's Orchard is in the center of the picture. Photo point looks southwest and is approximately 50 yards from the intersection. NATIONAL PARK SERVICE PHOTOGRAPH, PHOTOGRAPHER UNKNOWN

PLATE 10B. **June 1987 (30 years later)** These slopes were also a site of heavy tree removal during the dwarf mistletoe control program. In this area, however, most of the trees were poisoned and left standing. Their snags, stark and beautiful, persist today, a consequence of the slow decay rate of limber pine. They provide valuable nesting habitat for bluebirds. The apparent increase in the cover of perennial grasses in the foreground, which include bluebunch wheatgrass, probably reflects seasonal variations in moisture rather than a permanent change.

PLATE IIA. **Summer 1932?** View of the north slopes of North Crater taken from just west of the present parking area. NATIONAL PARK SERVICE PHOTOGRAPH, PHOTOGRAPHER UNKNOWN

PLATE IIB. **June 1990 (58 years later)** This was an area of selective removal of mature limber pine by poisoning. Many of the snags remain standing, and there has been substantial regeneration of new limber pine. The area in the foreground was used for camping in the 1930s but the area was closed probably when the first loop road was developed, and appears to have rehabilitated well.

PLATE 12A. **Summer 1952** The view is up Little Cottonwood Canyon in the northern portion of the monument. Photo point is on the slopes east and above the road up the canyon. Photos are not exact duplicates because the photo point was difficult to locate due to the loss of reference points and changes in aspen. The highest ridge in the left top center marks the northern boundary of the monument and is the highest point at 7,696 feet. The buildings in the foreground of the original photo are part of the Martin Mine. This small gold mine was active in the early part of the century and was acquired by the National Park Service in the 1950s. NATIONAL PARK SERVICE PHOTOGRAPH, PHOTOGRAPHER UNKNOWN

PLATE 12B. **June 1986 (34 years later)** All of the mine buildings have been eliminated and most sites restored. The remains of one small building in the center of Plate 12b have been burned since this photo was taken. A small pile of mine tailings from the main shaft, visible on the slope in the right center of the photo has remained totally devoid of any vegetation since it was deposited, probably in the 1920s. Soil tests have shown the tailings to be extremely high in toxic elements. There has been a noticeable decrease in aspen over the time period. This is the opposite of trends elsewhere in the monument which show a marked increase in aspen densities. The decline on this site is probably due to removal by beaver that became established after the mine operations ceased. There has been no apparent change in the sagebrush/steppe vegetation.

PLATE 13A. **Summer 1935** View looking south across Trench Mortar Flat from the bluff west of Crescent Butte. The prominent hill on the left is Coyote Butte, while the two-peaked butte to the right is The Watchman. The dotted line on the 1935 photo was made on the negative by engineer Scofield as the route of a proposed road to the southern portion of the monument. NATIONAL PARK SERVICE PHOTOGRAPH BY J. S. SCOFIELD

PLATE 13B. **June 1986 (51 years later)** The wilderness trail now follows the proposed road route through the southern end of the monument. Motorized access is now prevented in this portion of the monument due to its designation as a wilderness area in 1970. The most visible change in this photo is due to the fire which burned over Coyote Butte probably sometime in the 1940s. A relatively dense stand of perennial grasses, primarily Great Basin wild rye, now dominates the butte, which was formerly dominated by sagebrush. All of the limber pine on the butte appear to have been killed by the fire.

PLATE 14A. **May 1923** The camera faces southeast from about the 5,800-foot contour about a third of a mile down slope from the summit of Big Cinder Butte. It looks down on an unnamed cinder depression directly toward Coyote Butte in the center of the photo. PHOTOGRAPH BY ROBERT LIMBERT

PLATE 14B. **June 1986 (63 years later)** The impact of fire in removing the limber pine from Coyote Butte is clearer than in the previous plate. The density and size of the limber pine surrounding the cinder depression has increased markedly whereas the patterns of scattered shrubs and bare ground in the depression have not changed at all. Note the persistence of and similarity in appearance of the limber pine snag in the right foreground. The reason for the lack of vegetation in the cinder depression is unclear. No soil studies have been done on the area.

PLATE 15A. **May 1961** The camera faces the south side of Paisley Cone. The loop road passes through the photo in the middle. The camera point is on the 6,000-foot contour on the northeast side of Inferno Cone. NATIONAL PARK SERVICE PHOTOGRAPH BY DAVE OCHSNER

PLATE 15B. **June 1986 (25 years later)** The two scenes appear to be virtually identical. The patterns of the shrub communities, primarily antelope bitterbrush, rubber rabbitbrush, with some wax current and big sagebrush, appear to be unchanged reflecting the difficulty in colonizing barren cinder areas.

PLATE 16A. **Summer 1932?** This view, which was taken from the slopes above the road entering the monument, looks south-southwest. The largest log structure in the center of the 1930s scene is the Craters Inn, which held a restaurant, lodge, and some offices. The structure in the foreground was a gas station. The small log cabins on the left were rented out to tourists. Areas in the center foreground and to the left of the photo were used for camping. The 1986 photo shows the present entrance station. Except for two sites on the left of the road, camping is limited to the campground, which is to the right of the entrance station.   NATIONAL PARK SERVICE PHOTOGRAPH, PHOTOGRAPHER UNKNOWN

PLATE 16B. **May 1991 (59 years later)** The two scenes are quite comparable despite the fact that this area has been the site of relatively intense human activity. There has been some expansion of the shrub community, mostly antelope bitterbrush and rubber rabbitbrush, on the area occupied by buildings in the 1930s. These buildings were apparently removed in the late 1950s. The shrub patterns visible on Paisley Cone and North Crater in the background show little change. There has been a decrease in limber pine on the lower slopes of North Crater as a result of mistletoe control.

PLATE 17A. **Summer 1961** The camera position is just north of Highway 20 about 1/2 mile west of the monument entrance road. The view is to the northwest with the Pioneer Mountains in the background. The lobe of the Highway Lava Flow is on the left. NATIONAL PARK SERVICE PHOTOGRAPH, PHOTOGRAPHER UNKNOWN

PLATE 17B. **June 1986 (25 years later)** The two scenes are very comparable. The shrub communities dominated by antelope bitterbrush are little changed. The greater plant cover on the cinder gardens in the foreground probably reflects the fact that it was taken earlier in the spring when soil moisture was higher. Differences in lighting don't permit comparative observations on the plant communities in the Pioneer Mountains.

PLATE 18A. **Summer 1935** The camera faces southeast about 50 yards west of the wilderness trail. The view is of the lower slopes of Coyote Butte with Echo Crater in the background. NATIONAL PARK SERVICE PHOTOGRAPH BY J. S. SCOFIELD

PLATE 18B. **June 1988 (53 years later)** Indications are that this section was burned in the Coyote Butte fire in the 1940s, although the effects are not obvious today. Shrubs, mostly sagebrush and grass cover (Great Basin wildrye) are much as they appeared in the original scene.

PLATE 19A. **Summer 1955** A view to the northwest from the northwest slopes of Echo Crater. Big Cinder Butte is seen in the background with the Pioneer Mountains visible on the horizon of the 1986 photo. NATIONAL PARK SERVICE PHOTOGRAPH BY COLEMAN NEWMAN

PLATE 19B. **June 1986 (31 years later)** This comparison as much as any in this book shows the relatively constant character of the shrub communities and their inability to invade barren cinder gardens probably because of differences in soil characteristics.

PLATE 20A. **May 1923** The camera position is only a few yards to the right and down slope from that used in Plate 19. The view is almost the same. PHOTOGRAPH BY ROBERT LIMBERT

PLATE 20B. **May 1988 (65 years later)** This photo shows similar characteristics to those described in Plate 19. Haze from numerous spring fires in the region has obscured the Pioneer Mountains in the 1988 photo.

PLATE 21A. **June 1957** View looking almost due north to the south slopes of Half Cone. Photo point is on the wilderness trail in Trench Mortar Flats about 25 yards north of the point used for Plate 13. NATIONAL PARK SERVICE PHOTOGRAPH BY COLEMAN NEWMAN

PLATE 21B. **June 1988 (31 years later)** The remarkable constancy of shrub patterns with bare cinder areas shows well in the comparison. No comprehensive soil studies have been done in the monument. The limited number of soil pits dug in other parts of the monument suggest that the cause of such vegetation patterns are buried soil horizons. The large shrubs in the foreground are rubber rabbitbrush.

PLATE 22A. **May 1923** View looking south from the area just west of Trench Mortar Flats in the vicinity of Coyote Butte. Dead and down limber pines can be seen on the far right of the photo. The individual posing is part of the Limbert exploration party. The shadow in the lower left is probably that of Limbert. PHOTOGRAPH BY ROBERT LIMBERT

PLATE 22B. **May 1988 (65 years later)** Virtually no change is evident in the composition of the vegetation on this large cinder garden area. The vegetation consists of silverleaf phacelia and dwarf buckwheat. During the late spring, particularly when there has been abundant winter moisture, these cinder gardens are covered by the beautiful blooms of the purple-colored dwarf monkeyflower. The dead limber pines are still visible to the right of the photo.

PLATE 23A. **Summer 1935** The ridge of Echo Crater as seen from the sagebrush flats to the northeast. The rocky clefts and isolation of this crater make it an important nesting area for several species of raptors. NATIONAL PARK SERVICE PHOTOGRAPH BY J. S. SCOFIELD

PLATE 23B. **June 1986 (51 years later)** There is little apparent difference between the two photographs. The shrub communities appear to be of the same density and composition. For unknown reasons, there has been no invasion of limber pine into this mature sagebrush/steppe stand.

PLATE 24A. **May 1923** The camera position is on the south rim of Echo Crater, along the left portion of the rim shown in Plate 23. The view is to the southeast toward The Watchman. Individuals in the photo are part of the Limbert exploration party. This photo appeared in the 1924 issue of National Geographic. PHOTOGRAPH BY ROBERT LIMBERT

PLATE 24B. **May 1988 (65 years later)** The 1988 camera position was shifted slightly to the left because the view was blocked by the limber pine in the foreground. The relative stability of the plant communities is evident from the small amount of change which has occurred in the area. This is not unexpected as 65 years is a short time span for dramatic primary succession changes to have occurred.

PLATE 25A. **Summer 1935** The camera faces southeast with Crescent Butte visible on the horizon. The photo point is along the present wilderness trail between Broken Top and Big Cinder Butte, whose lower slopes are visible on the right. NATIONAL PARK SERVICE PHOTOGRAPH BY J. S. SCOFIELD

PLATE 25B. **June 1986 (51 years later)** Aside from an increase in the density of limber pine, there has been little change in the vegetation over the years. The rocky, coarse cinder areas in the foreground remain barren probably due as much to the southern exposure, lack of soil, and harsh environment as to any other factors. A source of seed for revegetation shouldn't be a problem. Over a longtime period these areas may eventually become vegetated. The dashed lines on the original indicate the route of the proposed road into the southern part of the monument.

PLATE 26A. **Summer 1935** This view, taken about 1,000 feet north of Plate 25 along the slopes of Broken Top, also looks to the south with Crescent Butte in the background. Although not clearly identifiable, Buffalo Cave is in the center of the photo. The dotted line on the 1935 photo was made on the negative by engineer Scofield as the route of a proposed road to the southern portion of the monument. NATIONAL PARK SERVICE PHOTOGRAPH BY J. S. SCOFIELD

PLATE 26B. **June 1986 (51 years later)** Plant succession on lava flows such as these is extremely slow. The most important factor determining vegetation growth on lava flows is the amount of soil which has accumulated. It appears that aa flows take longer for vegetation to become established because it takes more time for sufficient soil to accumulate and cover the irregular surfaces of these flows. Aa flows are shown on the left of this photo; the smooth pahoehoe flow is on the right.

PLATE 27A. **Summer 1935** The camera looks north from the lower slopes of Broken Top. The slope on the left of the photo is Inferno Cone. The trail in the original scene was the access into the tree molds area. The paved road in the 1988 scene follows roughly the same course. Duplicating the original camera position was difficult because of a lack of identifiable foreground characteristics. NATIONAL PARK SERVICE PHOTOGRAPH BY J. S. SCOFIELD

PLATE 27B. **June 1988 (53 years later)** This scene is another example of the difficulty vegetation has becoming established on the harsh aa flows. There is also no apparent change in the patterns of established vegetation in the background.

PLATE 28A. **Summer 1935** The camera faces northwest from the turnout below the tree molds parking area. The camera position in the 1986 scene is shifted slightly to the west. NATIONAL PARK SERVICE PHOTOGRAPH BY J. S. SCOFIELD

PLATE 28B. **June 1986 (51 years later)** No apparent change in vegetation can be seen on these harsh lava flows due to the slow pace of plant succession. The first plants to become established are lichens. As joints and crevices in the lava are filled with soil, the first vascular plants become established. These typically include lava phlox and dwarf goldenweed.

PLATE 29A. **Summer 1957** View to the southeast on the southeast slope of Sunset Cone at about the 6,100-foot contour. Scene shows the present park headquarters, visitor center, and maintenance and housing area. All are under construction in the original photo. NATIONAL PARK SERVICE PHOTOGRAPH BY COLEMAN NEWMAN

PLATE 29B. **June 1986 (29 years later)** The effects of the construction activities have been mitigated by site preparation including fertilization, which facilitated plant succession. Records indicate that no actual re-planting was carried out on the disturbed sites. At the time, the importance of re-planting disturbed areas with native vegetation was not widely recognized. The construction of facilities was part of a massive nation-wide National Park Service effort called Mission 66 designed to rebuild the infrastructure of the parks, neglected because of funding shortages caused by World War II. Plant communities visible in most other areas appear to be relatively unchanged, but there seems to be somewhat more limber pine on the slopes in the left foreground in the 1986 photo.

PLATE 30A. **Summer 1930?** This view is taken higher on the slope southwest of the current residential housing area than Plate 16 with the camera angle more to the southwest. It is more to the right than the camera location for Plate 31. The view is of the Crater's campground area as it existed in the 1930s and at present. Although the exact date is unknown, the crowds in the area suggest it was probably taken on a May second, celebrated for many years as Founder's Day, which was the date President Coolidge proclaimed the area as a national monument. NATIONAL PARK SERVICE PHOTOGRAPH, PHOTOGRAPHER UNKNOWN

PLATE 30B. **June 1989 (59 years later)** Most of the development seen in the original photo has been removed. The campground area remains, although now campsites are strictly defined. The two buildings at the right of the main road and the Craters Inn do not appear in the original photo from Plate 16 suggesting that this is a later photo. The large building with the chimney appears in Plate 32, although because of the different camera angle the resemblance is not clear.

PLATE 31A. **Summer 1935** This is another view of the Crater's developed area with the Craters Inn and tourist cabins on the left. The 1989 photo, which shows the current entrance kiosk, was taken at a lower elevation due to grading and slope changes associated with the development of the employee housing area. NATIONAL PARK SERVICE PHOTOGRAPH BY W. BICKNELL

PLATE 31B. **June 1989 (54 years later)** It was surprising to the authors that so little information other than these photos exists on the historic development of the monument. Because of a combination of isolation, and prior use, tourist facilities and administrative functions at the monument probably intermeshed in a manner that would not be permitted today. Likewise, lacking current cultural resource sensitivity, the historic buildings shown in the original were removed with no study or concern about their cultural significance. As a consequence, there is little documentation on when they were built or what they were used for.

PLATE 32A. **May 30, 1936** View of the Crater's campground area taken from near the present entrance station. This building, also shown in Plate 30, probably served as the administrative offices for the monument. Echoing the comments made on the preceding plate, little documentation exists for some of the early structures at the monument. NATIONAL PARK SERVICE PHOTOGRAPH BY K. DOLE

PLATE 32B. **May 30, 1986 (50 years later)** The monument campground now occupies this location. The 52-site campground has been carved out of the lava formations and presents an extraordinary setting and some unusual campsites. It is often full during the summer months.

PLATE 33A. **Summer 1935** The 1986 photo shows the route of the present trail into the wilderness section of the monument. The slope in the background is the south edge of Broken Top. The sign in the 1986 photo marks the boundary of the designated wilderness area. The trail branches just over the ridge in front of the sign. The left fork continues to the tree molds parking area; the other branch goes north to the loop road. NATIONAL PARK SERVICE PHOTOGRAPH BY J. S. SCOFIELD

PLATE 33B. **June 1986 (51 years later)** As in previous photos by Scofield, the dashed lines indicate the route of the proposed road into the southern section of the monument. The changes in the use and character of the monument would have been dramatic had this road been built. Although it might have stimulated greater visitor use, it also would have had greater impact on the resources and solitude the area now offers. It would also have precluded wilderness designation. There is no documentation on any follow-up to Scofield's study. It is probable that road-building plans ran into funding limitations imposed by World War II.

PLATE 34A. **June 1960** The camera faces south-southwest from the summit of Big Craters overlooking the two spatter cones. Big Cinder Butte is the largest mountain in the distance. NATIONAL PARK SERVICE PHOTOGRAPH BY DAVE OCHSNER

PLATE 34B. **June 1986 (26 years later)** The spatter cones are unique features because of their structural integrity and depth. The Limbert exploration commented that they dropped a rock in them and couldn't hear it touch bottom, although in actuality all are less than 100 feet deep. Because of their location near the primary use areas, the cones have received extensive visitor use which has resulted in resource damage. The trails and resulting erosion around and over the spatter cones common in earlier years are evident in the original photo. In recent years, more restrictive visitor use policies combined with efforts to restore resources to their natural condition have helped heal some of the impacts of the past.

What the Photo Comparisons Reveal

The photographs reveal the great variations in the rate that successional changes are occurring within the vegetation types in the monument. They also indicate that both natural and human-caused disturbance of the vegetation has been very localized and most of the vegetational changes are related to the process of primary succession. Primary succession is the mechanism by which changes in plant communities occur on a site over time as soil and other factors develop. The rate at which fertile, mature soils—that is, soils with a substantial amount of organic matter and moisture-holding capacity—develop from the sterile lava flows and cinder material is probably the main factor limiting the development of vegetation.

An examination of the older lava flows on the monument has indicated that the accumulation of loess—very fine wind-deposited soil—has had a profound influence on the soil development process. Soil development on the lava flows can also occur through chemical and physical weathering of the lava, which slowly breaks it down into smaller particles capable of supporting plant growth and matures it through the slow accumulation of organic material. The accumulation of a fertile soil layer proceeds at a slower rate on the aa flows than on the pahoehoe flows because the irregular surface of the aa lava flows requires a greater accumulation of fine soil material to support plant growth (Eggler 1941). As a result of the slow development of soil profiles on these flows, changes in plant communities occur very slowly and the composition of the lava-dominated communities has remained extremely stable throughout the period (see photo numbers 26, 27, 28, and 29).

The composition of the vegetation type we have termed cinder gardens—that is, vegetation growing on the loose cinder soil layers—has also been stable for a long time as evidenced by the photographs (see photo numbers 2, 15, 17, 21 and 22). The development of mature organic soils from cinder material is extremely slow, and the establishment of many species is probably limited by the fact that most sites still have immature soils. Cross-sections through the cinder material often indicate the presence of older buried soils. Buried soils slowly developed and matured following a volcanic eruption and were then covered by cinders from a more recent eruption. In some areas several buried soil layers may be found. In most cases the most recent surface soils tend to dictate the kind of vegetation that occurs on a site. However, in some cases where mature buried soil layers lie close to the surface, plants with deeper root systems will take hold. This phenomenon helps explain the often disjunct distribution and sharp boundaries of some plant communities.

The more recent cinder deposits are occupied by two low growing perennials, dwarf buckwheat and silverleaf phacelia. The delicate annual dwarf monkeyflower may be briefly common in spring, and in particularly moist years may blanket the cinder areas with an astonishing array of colors. This display is short-lived, however, and the blossoms and plants soon wither and disappear with the advent of early summer heat, which is intensified by the black soil surface of the cinders.

Antelope bitterbrush and rubber rabbitbrush are among the first shrubs to become established as fine material accumulates through the soil development process. Establishment of plant seedlings in the cinder material is an extremely difficult process because the cinders are low in nutrients and retain little moisture. Plants often become established in the depressions in the surface that result from tracks left by animals and vehicles because these offer a more favorable moisture and nutrient environment for the seedlings (Limbert 1924, Eggler 1941, Day 1985). It is ironic that despite the fact that primary succession occurs slowly on the cinder gardens, these sites recover surprisingly quickly following disturbance (see photo number 11).

The composition of most plant communities dominated by antelope bitterbrush has remained relatively stable over long time periods (see photo numbers 15, 17, 19, and 21), while others have shown marked increases in limber pine (see photo numbers 1, 2, and 5). This difference may be due to a number of ecological factors. Antelope bitterbrush plant communities form an intermediate successional stage between the dwarf buckwheat and silverleaf phacelia plant communities that initially become established on the cinder cones and the limber pine communities that eventually become established on the older cinder material. It seems clear that the rate and degree of soil development plays an essential role in determining how long the antelope bitterbrush communities persist and the rate at which limber pine establishment occurs (Day 1985, Eggler 1941).

The availability of propagules or seeds may also be an important factor in plant establishment, and in the case of limber pine a resident bird species, the Clark's nutcracker, plays an essential role in its regeneration by transporting and burying seeds in the soil (Lanner and Van DerWall 1980). As a consequence of this efficient transport mechanism, the immediate availability of seeds is less critical for limber pine than most other species.

Antelope bitterbrush reproduction is also enhanced through animal seed caching activity. For this species, the animals involved are primarily rodents. Abandoned rodent caches provide a more suitable site for seedling establishment than seeds which are simply broadcast on the soil surface. The rodents may also aid in seed dispersal, albeit on a smaller scale than that provided by Clark's nutcrackers (Evans et al. 1983).

The most dramatic changes in the vegetation on the monument which have occurred in the time frame we have investigated can be observed in the conifer-dominated vegetation types and the sagebrush/steppe adjacent to them. There has been a general increase in limber pine throughout the northern part of the monument (see photo numbers 1, 2, 3, 4, and 5). The increased dominance of limber pine can be observed in many areas except where the dwarf mistletoe control program was conducted in the early 1960s (see photo numbers 8, 9, and 10). On several sites the removal of mistletoe-infected trees appears to have been compensated for by the establishment of young, new trees (see photo number 11).

Douglas-fir trees have also increased in density and have slowly spread into adjacent sagebrush/steppe vegetation (see photo numbers 6 and 7). The increasing density of the conifer tree canopy, which over time alters sunlight and moisture, has substantially changed the shrub understory composition of these areas. Remnant plants and skeletons of shrubs typical of sagebrush/steppe can now be found in the tree stands. Antelope bitterbrush and big sagebrush are sensitive to shading and will decline in vigor and reproductive rate as the conifer overstory closes. As this trend continues, the composition of the vegetation will completely change from a sagebrush/steppe vegetation toward a forested type.

The increase of limber pine and Douglas-fir on the monument is consistent with observation of the general increase in conifers throughout southern Idaho and the northern Great Basin, as documented by several studies (Loope and Gruell 1973, Burkhardt and Tisdale 1976, Dunwiddie 1977, Arno 1980, Arno and Gruell 1983, Gruell 1983, Butler 1986). What is interesting is that the increase in conifers at Craters of the Moon may be due to different factors than have been hypothesized elsewhere in the region. Most of the above studies have proposed that the principal factor influencing the increase in conifers has been a reduction in fires as a result of human management. However, due to the sparse plant cover and broken topography of the monument, fire has not been prevalent and has probably not played a major role in the increase of conifers there. Studies we have conducted on the monument (Bunting and Wright 1991) have clearly indicated that the extent of individual fires has been severely limited, although we have found charcoal and evidence of other small ignitions, such as fire scars on trees and the charred roots of shrubs, in nearly all vegetated areas on the monument. Most of these ignitions covered very small areas because of the factors alluded to above. The photographs document only a single fire occurrence, the fire on Coyote Butte in the 1940s (see photo numbers 13, 14, and 18). Assuming that fire was an uncommon event, fire control probably has not affected the increase in conifers and other successional changes on the monument to the degree that has been reported elsewhere. What has caused the changes in conifer composition seen in the photographs? We speculate that perhaps climatic variation has created a more favorable environment for conifer establishment and has been more of a factor in this instance than the reduction in fire occurrence. However, an examination of precipitation and temperature records from the monument have shown no clear correlations.

Limber pine stands in the probably drier central portion of the monument have been more stable. In these areas, there appears to have been little change in tree density or distribution over the time period we have been able to observe (see photo numbers 25, 27, 28 and 33). We speculate that these trees are probably growing near the environmental tolerance limit for limber pine, and available moisture probably limits the establishment of pine to only the most favorable sites. Even then, pine reproduction may occur only during the most favorable years.

Changes that have occurred in the aspen-dominated vegetation over the period presented in the photographs are variable. There was a decline in the aspen communities in the upper portion of Little Cottonwood Creek (see photo number 12). This may have been attributable to the occasional use of the area by beaver. Other sites we have investigated, which are not included in the photographs, indicate that Douglas-fir has invaded into the stands and there aspen are also declining. In still other areas it appears from the size-class distribution of the tree trunks that aspen is advancing into sagebrush/steppe vegetation. Unfortunately aspen is not well represented in the photographs, making conclusive statements difficult. Throughout southern Idaho and the northern Great Basin there has been an invasion into aspen stands by conifers, particularly Douglas-fir and juniper, during the past century (Loope and Gruell 1973, Gruell 1983).

The sagebrush/steppe vegetation within the monument is not extensive in terms of the area covered, but what does occur provides some interesting insights into the ecology of the vegetation type. Two major influences on sagebrush/steppe vegetation on the Snake River Plain have been grazing by domestic livestock and fire. As stated, we have not found fire to have been a major influence during the past 100 years on the monument. We do know that the northern portion of the monument was grazed by sheep and cattle from the 1870s until the land was placed under the jurisdiction of the National Park Service in 1924. In the central and southern portions of the monument inaccessibility due to rough lava flows has always minimized the influence of livestock grazing. This is not to say there weren't efforts to use this range. For example, there is an old stock water tank constructed from lava rock in the southern portion of the monument that was filled with buckets from one of the permanent water holes along the Great Rift. It is doubtful however that this operation persisted from many years.

Most investigators have concluded that the kipukas, such as Carey Kipuka and Round Knoll, have always been relatively free of grazing and fire. Because Craters of the Moon has had few of these influences, the monument provides a unique opportunity to observe ecological relationships in the sagebrush/steppe vegetation type, unlike most other areas on the Snake River Plain.

Data from the kipukas (Tisdale et al. 1965) and photographic evidence from the northern portion of the monument (see photo numbers 6, 7, 12 and 29), which has not been influenced by livestock for over 50 years, indicate that sagebrush is an important component of the climax vegetation and that a grassland will not develop in the absence of fire, as suggested by some authors. A wildfire occurred in 1985 immediately north of the monument. The authors established photographic points and have collected quantitative data on plant composition changes since that time. This information will provide an opportunity to study the vegetation changes initiated by fire.

Observations of different age volcanic deposits and of the Coyote Butte fire (see photo numbers 13, 14, and 18) indicate that big sagebrush becomes established more slowly than other shrubs, such as antelope bitterbrush and rubber rabbitbrush, during both primary and secondary succession. The factor(s) which limit the establishment of sagebrush are not clear, however. We do know that once established, sagebrush becomes a dominant component of the plant community and, with no disturbance, the density of plants remains relatively stable over long periods of time. Individual plants may readily live for over 75 years and many have survived for over 100 years. As with many other plant species, those sagebrush growing on severe sites may have the longest lifespan. We have observed a decline in sagebrush only in areas where conifer establishment has significantly altered the environment of a site.

A number of common non-native species are present in various plant communities found on the monument, excluding those species found in the immediate vicinity of the visitor center. The most widespread of these exotic species are cheatgrass, salsify, mullein, and Russian thistle. Mullein and Russian thistle occur mainly as roadside weeds and have spread a short distance into adjacent vegetation. Cheatgrass and salsify are ubiquitous in the monument, although they have not had an obvious affect on the plant communities in which they are present. Cheatgrass does have the potential of altering the fire regime and perennial grass seedling establishment particularly on the kipukas and areas with more advanced soil development. Sites with more developed soils support a more productive vegetal growth and therefore have a greater potential to burn. Another exotic plant, Canada thistle, occurs primarily along permanent and intermittent water courses and is abundant in the Great Basin wildrye vegetation in the northern portion of the monument. Its presence affects these communities, but the potential for spread is limited by moisture.

Putting together the story of vegetation change at Craters of the Moon from photographic evidence has been a rewarding and challenging experience. The photos have forced us to ask a lot of questions about why things are as they appear. In the process, we have learned a great deal about the ecological processes that have occurred and are on-going on the monument. We hope through this book to share some of these insights with others. We have been challenged by a great number of questions to which we don't have answers.

It is probably natural that a study such as this would at its conclusion pose more questions than answers. We were constrained by the number and quality of historic photos we were able to locate. Almost all of these photos were taken for a specific reason, but most were not taken with the viewpoint that they would be valuable in the study of vegetation change years later. For example, we found no historic close-up photographs which would show more detail of the herbaceous cover. We also wish we had more photographs of some areas, such as the asper groves in the north end and of the kipukas. There maybe in fact, photographs of some of the areas we have not yet found but which will be made known to us by readers of this book.

These factors have led us to establish a series of photographic points throughout the monument which are situated expressly to examine long-term changes in environmental conditions. Perhaps if we—or investigators not yet born—are able to retake photos in future years, this information will further document, interpret, and add to our understanding of the ecological processes at work in this fascinating landscape we have been privileged to work in.

Appendix: Common and Scientific Names of Plants

Common Name	Scientific Name	Common Name	Scientific Name
bluebunch wheatgrass	*Agropyron spicatum*	lava phlox	*Leptodactylon pungens*
dwarf mistletoe	*Arceuthobium campylopodum*	dwarf monkeyflower	*Mimulus nanus*
big sagebrush	*Artemisia tridentata*	silverleaf phacelia	*Phacelia hastata*
cheatgrass	*Bromus tectorum*	limber pine	*Pinus flexilis*
rubber rabbitbrush	*Chrysothamnus nauseosus*	quaking aspen	*Populus tremuloides*
Canada thistle	*Cirsium arvense*	Douglas-fir	*Pseudotsuga menziesii*
Great Basin wildrye	*Elymus cinereus*	antelope bitterbrush	*Purshia tridentata*
dwarf buckwheat	*Eriogonum ovalifolium*	wax current	*Ribes cereum*
Idaho fescue	*Festuca idahoensis*	Russian thistle	*Salsola iberica*
dwarf goldenweed	*Haplopappus nanus*	salsify	*Trgopogon dubius*
Rocky Mountain juniper	*Juniperus scopulorum*	mullein	*Verbascum thapsus*

Literature Cited

Arno, S. F. 1980. Forest fire history in the northern Rockies. Journal of Forestry 78:460-465.

Arno, S. F., and G. E. Gruell. 1983. Fire history at the forest-grassland ecotone in southwestern Montana. Journal Range Management 36:332-336.

Blakesley, J. A., and R. G. Wright. 1988. A review of scientific research at Craters of the Moon National Monument. University of Idaho Forest, Wildlife and Range Experiment Station Bulletin 50. 41 p.

Bunting, S. C., and R. G. Wright. 1991. A fire management plan for Craters of the Moon National Monument. Unpubl. Report University of Idaho. CPSU.

Burkhardt, J. W., and E. W. Tisdale. 1976. Causes of juniper invasion in southwestern Idaho. Ecology 57:472-484.

Butler, D. R. 1986. Conifer invasion of subalpine meadows, central Lemhi Mountains, Idaho. Northwest Science 60:166-173.

Christensen, E. M. 1957. Photographic history of the mountain brush on "Y" Mountain, central Utah. Utah Academy of Sciences, Arts and Letters, Proceedings 34:154-155.

Day, T. A. 1985. Plant association and soil factors in primary succession on cinder cones in Idaho. M.S. Thesis. University of Idaho. Moscow. 62 p.

Day, T. A., and R. G. Wright. 1985. The vegetation types of Craters of the Moon National Monument. University of Idaho Forest, Wildlife and Range Experiment Station Bulletin 38. 6 p.

Day, T. A. and R. G. Wright. 1989. Positive plant association with Eriogonum ovalifolium in primary succession. Vegetatio 80:37-45.

Dunwiddie, P. W. 1977. Recent tree invasion of subalpine meadows in the Wind River Mountains, Wyoming. Arctic and Alpine Research 9:393-399.

Eggler, W. A. 1941. Primary succession of volcanic deposits in southern Idaho. Ecological Monographs 3:277-298.

Evans, R. A.; J. A. Young; G. J. Cluff; and J. K. McAdoo. 1983. Dynamics of antelope bitterbrush seed caches. pages 195-202. IN A. R. Tiedeman and K. J. Johnson, compilers. Proceedings Research and management of bitterbrush and cliff rose in western North America. U. S. Forest Service General Technical Report INT-152.

Griffith, B. 1983. Ecological characteristics of mule deer at Craters of the Moon National Monument. University of Idaho CPSU Report B-83-2. 109p.

Gruell, G. E. 1980. Fires influence on wildlife habitat on the Bridger-Teton National Forest, Wyoming. Volume I—Photographic record and analysis. U.S. Forest Service Research Paper INT- 235. 207 p.

Gruell, G. E. 1983. Fire and vegetative trends in the northern Rockies: Interpretations from 1871-1982 photographs. U. S. Forest Service General Technical Report INT-158. 117 p.

Harrison, A. E. 1974. Reoccupying unmarked camera stations for geological observations. Geology 2:469-471.

Hastings, J. R., and R. M. Turner. 1965. The changing mile. University of Arizona Press. Tucson. 317 p.

Hironaka, M.; M. A. Fosberg; and A. H. Winward. 1983. Sagebrush-grass habitat types of southern Idaho. University of Idaho Forest, Wildlife and Range Experiment Station Bulletin 35. 44 p.

Houston, D. B. 1982. The northern Yellowstone elk. Macmillan Publ. Co. New York. 473 p.

Irving, W. 1868. The adventures of Captain Bonneville, U. S. Army. Hudson Co. New York.

Kuntz, M. A.; D. E. Champion; E. C. Spiker; and R. H. Lefebvre. 1986a. Contrasting magma types and steady-state, volume-predictable, basaltic volcanism along the Great Rift, Idaho. Bulletin Geological Society of America 97:579-594.

Kuntz, M. A.; E. C. Spiker; M. Rubin; D. E. Champion; and R. H. Lefebvre.

1986b. Radiocarbon studies of latest Pleistocene and Holocene lava flows of the Snake River Plain, Idaho. Quaternary Research 25:163-176.

Lanner, R. M., and S. B. Van DerWall. 1980. Dispersal of limber pine seed by Clark's nutcracker. Journal Forestry 78:636-639.

Limbert, R. W. 1924. Among the Craters of the Moon. National Geographic Magazine 45:303-328.

Loope, L. L. and G. E. Gruell. 1973. The ecological role of fire in the Jackson Hole area, northwestern Wyoming. Quaternary Research 3:425-443.

Mack, A.; W. P. Gregg; S. P. Bratton; and P. S. White. 1983. A survey of ecological inventory, monitoring, and research in the U. S. National Park Service Biosphere Reserves. Biological Conservation 26:33-45.

Mack, R. N. 1981. Invasion of Bromus tectorum L. into western North America: An ecological chronicle. Agro-Ecosystems 7:145-165.

Magnuson, J. J. 1990. Long-term ecological research and the invisible present. BioScience 40:495-501.

Ostrogorsky, M. 1983. Historical overview for the Craters of the Moon National Monument of Idaho. Unpubl. Rep. University of Idaho CPSU. 28p.

Prinz, M. 1970. Idaho rift system, Snake River Plain, Idaho. Geological Society of America Bulletin 81:941-948.

Rees, R. A., and A. Sandy (eds.). 1977. The adventures of Captain Bonneville. Twain Publishers, Boston, MA. 346 p.

Rogers, G. F. 1982. Then and now. University of Utah Press, Salt Lake City. 152 p.

Russell, I. C. 1902. Geology and water resources of the Snake River Plain in Idaho. U. S. Geological Survey Bulletin 199:1-192.

Silsbee, D. G., and D. L. Peterson. 1991. Designing and implementing comprehensive long-term inventory and monitoring programs for National Park System lands. Natural Resource Report NPS/NRUW/NRP-91/04.

Stearns, H. T. 1924. Craters of the Moon National Monument. Geography Review 14:362-372.

Stearns, H. T. 1928. A guide to the Craters of the Moon National Monument Idaho. Bulletin Idaho Bureau of Mines and Geology. Vol 13. 59p.

Stearns, H. T. 1963. Geology of Craters of the Moon Idaho. Craters of the Moon Natural History Association. Arco, Idaho. 34 p.

Steele, R.; R. D. Pfister; R. A. Ryker; and J. A. Kittams. 1981. Forest habitat types of central Idaho. U. S. Forest Service General Technical Report INT-114. 138 p.

Swanson, F. J., and R. E. Sparks. 1990. Long-term ecological research and the invisible place. BioScience 40:502-508.

Tisdale, E. W.; M. Hironaka; and M. A. Fosberg. 1965. An area of pristine vegetation in the Craters of the Moon National Monument. Ecology 46:349-352.

Wright, R. G. 1992. Wildlife research and management in the national parks. University of Illinois Press, Urbana. 224 p.

Index

A

Aa lava, 5, 6, 7, 26, 32, 73, 75, 91
Antelope bitterbrush, 7, 23, 25, 51, 53, 55, 92, 93, 94
Aspen, 7, 21, 44, 45, 93, 94, 95

B

Basalt, 5, 6, 22
Baseline measurement, 1, 3, 13, 95
Beaver 45, 94
Big Cinder Butte, 5, 6, 22, 23, 28, 29, 48, 58, 70, 88
Big Craters, 5, 88
Bluebirds, 41
Bluebunch wheatgrass, 7, 41,
Bonneville, B. L. E., 2, 3
Broken Top, 70, 72, 74, 86
Buffalo Cave, 72

C

Campground, 8, 52, 80, 81, 84, 85
Carey Kipuka, 3, 94
Caves, 6, 8, 72
Cheatgrass, 19, 20, 95
Cinder cones, 6, 8, 22
Cinder garden, 21, 25, 31, 35, 55, 59, 65, 91
Clark's nutcracker, 92
Climate change, 2
Coyote Butte, 46, 47, 48, 49, 56, 57, 64, 93, 94
Craters Inn, 52, 81, 82

Crescent Butte 30, 31, 46, 70, 72

D

Devil's Orchard, 40
Douglas-fir, 7, 21, 33, 35, 92, 93, 94
Dwarf buckwheat, 25, 31, 65, 91, 92
Dwarf mistletoe (control), 20, 21, 37, 39, 41, 43, 53, 92
Dwarf monkeyflower, 65, 91

E

Echo Crater, 56, 58, 66, 68
Ecological change, 1, 91, 95
Exotic species, 2, 3, 19, 95

F

Fire, 1, 3, 19, 20, 21, 47, 49, 61, 93, 94, 95

G

Geology 5, 14,
Goodale's Cutoff, 3,
Grassy Cone 32, 33, 34
Grazing (domestic), 1, 3, 7, 19, 20, 94
Great Basin wildrye, 47, 57, 95
Great Owl Cavern, 6
Great Rift, 3, 5, 28, 94

H

Highway Flow, 32, 33, 54

I

Idaho fescue, 7,
Indian Tunnel, 6,
Inferno Cone, 24, 50, 74

K

Kipuka, 94, 95

L

Lava (lava flows), 2, 3, 5, 6, 7, 21, 26, 73, 77, 91
Lava domes, 6,
Lichens, 77
Limber pine, 7, 21, 23, 25, 27, 31, 37, 39, 41, 43, 47, 49, 53, 64, 65, 67, 68, 71, 79, 92, 93
Limbert, R., 3, 14, 22, 30, 64, 68, 89
Little Cottonwood Creek, 44, 93
Loess, 6. See also Soil(s)
Loop road, 8, 20, 36, 38, 40, 43, 50,

M

Martin mine, 44,
Mission 66 Program, 79
Monitoring (ecological), vii, 11, 12, 13
Mule deer, 7, 31,

N

Newman, Coleman, 14,
North Crater, 36, 37, 38, 42, 53
North Flow, 32

O

Oregon Trail, 3
Ostrogorsky, Michael, 3, 14

P

Pahoehoe flow, 5, 6, 27, 73, 91
Paisley Cone, 40, 50, 53
Pioneer Mountains, 3, 7, 26, 54, 55, 58, 61
Pollution, 2
Precipitation, 7

R

Repeat photography, 1, 13, 14
Residuum, 6. See also Soil(s)
Riparian vegetation 7
Rubber rabbitbrush, 7, 23, 51, 53, 63, 92, 94
Russell, I. C., 3, 5

S

Sagebrush 7, 19, 23, 33, 47, 51, 57, 66, 93, 94, 95
Sagebrush/steppe vegetation, 7, 19, 33, 35, 45, 67, 92, 93, 94
Silverleaf phacelia, 25, 65, 91, 92
Snake River Plain, vii, 1, 2, 3, 5, 7, 19, 94
Soil(s), 6, 7, 29, 35, 45, 59, 71, 77, 91, 92;
    buried horizons, 6, 25, 63, 73, 91;
    moisture in, 31, 55
Spatter cones, 6, 8, 88, 89
Stearns, H. T., 3, 5, 6
Succession (primary), 7, 69, 73, 77, 79, 91, 92, 94
Sunset Cone, 32, 78

T

Tree molds, 24, 26, 28, 74, 76, 86
Trench Mortar Flats, 28. 30, 46, 62, 64

V

Visitation (to Craters), 8, 20
Volcanic eruption 2, 5, 6, 11, 91

W

Washington, Irving, 2, 11
Wax current, 7, 51
Wilderness (area or designation), 2, 7, 14, 20, 47, 86, 87
Wilderness trail, 28, 30, 56, 62, 70, 86

Y

Yellowstone National Park, 13

*762-4
5-13
CC

DATE DUE

DUE DATE SUBJECT TO CHANGE
IF A RECALL IS REQUESTED

DEMCO, INC. 38-2931